INTRODUCTORY MESSAGE

Today, our country is on the verge of one of the most exciting and important innovations in transportation history—the development of Automated Driving Systems (ADSs), commonly referred to as automated or self-driving vehicles.

The future of this new technology is so full of promise. It's a future where vehicles increasingly help drivers avoid crashes. It's a future where the time spent commuting is dramatically reduced, and where millions more—including the elderly and people with disabilities—gain access to the freedom of the open road. And, especially important, it's a future where highway fatalities and injuries are significantly reduced.

Since the Department of Transportation was established in 1966, there have been more than 2.2 million motor-vehicle-related fatalities in the United States. In addition, after decades of decline, motor vehicle fatalities spiked by more than 7.2 percent in 2015, the largest single-year increase since 1966. The major factor in 94 percent of all fatal crashes is human error. So ADSs have the potential to significantly reduce highway fatalities by addressing the root cause of these tragic crashes.

The U.S. Department of Transportation has a role to play in building and shaping this future by developing a regulatory framework that encourages, rather than hampers, the safe development, testing and deployment of automated vehicle technology.

Secretary Elaine L. Chao
U.S. Department of Transportation

Accordingly, the Department is releasing *A Vision for Safety* to promote improvements in safety, mobility, and efficiency through ADSs.

A Vision for Safety replaces the Federal Automated Vehicle Policy released in 2016. This updated policy framework offers a path forward for the safe deployment of automated vehicles by:

- Encouraging new entrants and ideas that deliver safer vehicles;
- Making Department regulatory processes more nimble to help match the pace of private sector innovation; and
- Supporting industry innovation and encouraging open communication with the public and with stakeholders.

Thanks to a convergence of technological advances, the promise of safer automated driving systems is closer to becoming a reality. From reducing crash-related deaths and injuries, to improving access to transportation, to reducing traffic congestion and vehicle emissions, automated vehicles hold significant potential to increase productivity and improve the quality of life for millions of people. *A Vision for Safety* seeks to facilitate the integration of ADS technology by helping to ensure its safe testing and deployment, as well as encouraging the development of systems that guard against cyber-attacks and protect consumer privacy.

Our goal at the Department of Transportation is to be good stewards of the future by helping to usher in this new era of transportation innovation and safety, and ensuring that our country remains a global leader in autonomous vehicle technology.

EXECUTIVE SUMMARY

The world is facing an unprecedented emergence of automation technologies. In the transportation sector, where 9 out of 10 serious roadway crashes occur due to human behavior, automated vehicle technologies possess the potential to save thousands of lives, as well as reduce congestion, enhance mobility, and improve productivity. The Federal Government wants to ensure it does not impede progress with unnecessary or unintended barriers to innovation. Safety remains the number one priority for the U.S. Department of Transportation (DOT) and is the specific focus of the National Highway Traffic Safety Administration (NHTSA).

NHTSA's mission is to save lives, prevent injuries, and reduce the economic costs of roadway crashes through education, research, safety standards, and enforcement activity. As automated vehicle technologies advance, they have the potential to dramatically reduce the loss of life each day in roadway crashes. To support industry innovators and States in the deployment of this technology, while informing and educating the public, and improving roadway safety through the safe introduction of the technology, NHTSA presents *Automated Driving Systems: A Vision for Safety*. It is an important part of DOT's multimodal efforts to support the safe introduction of automation technologies.

In this document, NHTSA offers a nonregulatory approach to automated vehicle technology safety. *Section 1: Voluntary Guidance for Automated Driving Systems (Voluntary Guidance)* supports the automotive industry and other key stakeholders as they consider and design best practices for the testing and safe deployment of Automated Driving Systems (ADSs - SAE Automation Levels 3 through 5 – Conditional, High, and Full Automation Systems). It contains 12 priority safety design elements for consideration, including vehicle cybersecurity, human machine interface, crashworthiness, consumer education and training, and post-crash ADS behavior.

Given the developing state of the technology, this *Voluntary Guidance* provides a flexible framework for industry to use in choosing how to address a given safety design element. In addition, to help support public trust and confidence, the *Voluntary Guidance* encourages entities engaged in testing and deployment to publicly disclose Voluntary Safety Self-Assessments of their systems in order to demonstrate their varied approaches to achieving safety.

Vehicles operating on public roads are subject to both Federal and State jurisdiction, and States are beginning to draft legislation to safely deploy emerging ADSs. To support the State work, NHTSA offers *Section 2: Technical Assistance to States, Best Practices for Legislatures Regarding Automated Driving Systems (Best Practices)*. The section clarifies and delineates Federal and State roles in the regulation of ADSs. NHTSA remains responsible for regulating the safety design and performance aspects of motor vehicles and motor vehicle equipment; States continue to be responsible for regulating the human driver and vehicle operations.

The section also provides *Best Practices for Legislatures*, which incorporates common safety-related components and significant elements regarding ADSs that States should consider incorporating in legislation. In addition, the section provides *Best Practices for State Highway Safety Officials*, which offers a framework for States to develop procedures and conditions for ADSs' safe operation on public roadways. It includes considerations in such areas as applications and permissions to test, registration and titling, working with public safety officials, and liability and insurance.

Together, the *Voluntary Guidance* and *Best Practices* sections serve to support industry, Government officials, safety advocates, and the public. As our Nation and the world embrace technological advances in motor vehicle transportation through ADSs, safety must remain the top priority.

Over the coming months and years, NHTSA, along with other Federal agencies, where relevant, will continue to take a leadership role in encouraging the safe introduction of automated vehicle technologies into the motor vehicle fleet and on public roadways in the areas of policy, research, safety standards, freight and commercial use, infrastructure, and mass transit.

The **Office of the Under Secretary for Policy (OST-P)** is the office responsible for serving as a principal advisor to the Secretary and provides leadership in the development of policies for the Department, generating proposals and providing advice regarding legislative and regulatory initiatives across all modes of transportation. The Under Secretary coordinates the Department's budget development and policy development functions. The Under Secretary also directs transportation policy development and works to ensure that the Nation's transportation resources function as an integrated national system.
See www.transportation.gov/policy.

The **Office of the Assistant Secretary for Research and Technology (OST-R)** is the lead office responsible for coordinating DOT's research and for sharing advanced technologies with the transportation system. Technical and policy research on these technologies occurs through the Intelligent Transportation Systems (ITS) Research Program, the University Transportation Centers, and the Volpe National Transportation Research Center, which make investments in technology initiatives, exploratory studies, pilot deployment programs and evaluations in intelligent vehicles, infrastructure, and multi-modal systems.
See www.its.dot.gov and www.transportation.gov/research-technology.

The **Federal Motor Carrier Safety Administration (FMCSA)** is the lead Federal Government agency responsible for regulating and providing operational safety oversight (for instance, hours of service regulations, drug and alcohol testing, hazardous materials safety, vehicle inspections) for motor carriers operating commercial motor vehicles (CMVs), such as trucks and buses, and CMV drivers. FMCSA partners with industry, safety advocates, and State and local governments to keep our Nation's roadways safe and improve CMV safety through financial assistance, regulation, education, enforcement, research, and technology.
See www.fmcsa.dot.gov.

The **Federal Highway Administration (FHWA)** supports State and local governments in the design, construction, and maintenance of the Nation's highway system (Federal Aid Highway Program) and various Federal and tribal lands (Federal Lands Highway Program). Through financial and technical assistance to State and local governments, FHWA is responsible for ensuring that America's roads and highways continue to be among the safest and most technologically sound in the world.
See www.fhwa.dot.gov.

The **Federal Transit Administration (FTA)** provides financial and technical assistance to local public transit systems, including buses, subways, light rail, commuter rail, trolleys, and ferries. FTA also oversees safety measures and helps develop next-generation technology research.
See www.transit.dot.gov.

TABLE OF CONTENTS

Section 1: Voluntary Guidance

    Overview 1

    Scope and Purpose 2

    ADS Safety Elements 5

        System Safety 5

        Operational Design Domain 6

        Object and Event Detection and Response 7

        Fallback (Minimal Risk Condition) 8

        Validation Methods 9

        Human Machine Interface 10

        Vehicle Cybersecurity 11

        Crashworthiness 12

        Post-Crash ADS Behavior 13

        Data Recording 14

        Consumer Education and Training 15

        Federal, State, and Local Laws 15

    Voluntary Safety Self-Assessment 16

Section 2: Technical Assistance to States

    Overview 19

    Federal and State Regulatory Roles 20

    Best Practices for Legislatures 21

    Best Practices for State Highway Safety Officials ... 22

Conclusion 25

Endnotes 26

SECTION 1: VOLUNTARY GUIDANCE
For Automated Driving Systems

OVERVIEW

The U.S. Department of Transportation (DOT) through the National Highway Traffic Safety Administration (NHTSA) is fully committed to reaching an era of crash-free roadways through deployment of innovative lifesaving technologies. Recent negative trends in automotive crashes underscore the urgency to develop and deploy lifesaving technologies that can dramatically decrease the number of fatalities and injuries on our Nation's roadways. NHTSA believes that Automated Driving Systems (ADSs), including those contemplating no driver at all, have the potential to significantly improve roadway safety in the United States.

The purpose of this Voluntary Guidance is to support the automotive industry, the States, and other key stakeholders as they consider and design best practices relative to the testing and deployment of automated vehicle technologies. It updates the Federal Automated Vehicles Policy released in September 2016 and serves as NHTSA's current operating guidance for ADSs.

The Voluntary Guidance contains 12 priority safety design elements.[1] These elements were selected based on research conducted by the Transportation Research Board (TRB), universities, and NHTSA. Each element contains safety goals and approaches that could be used to achieve those safety goals. Entities are encouraged to consider each safety element in the design of their systems and have a self-documented process for assessment, testing, and validation of the various elements. As automated driving technologies evolve at a rapid pace, no single standard exists by which an entity's methods of considering a safety design element can be measured. Each entity is free to be creative and innovative when developing the best method for its system to appropriately mitigate the safety risks associated with their approach.

In addition, to help support public trust and confidence in the safety of ADSs, this Voluntary Guidance encourages entities to disclose Voluntary Safety Self-Assessments demonstrating their varied approaches to achieving safety in the testing and deployment of ADSs.[2]

Entities are encouraged to begin using this Voluntary Guidance on the date of its publication. NHTSA plans to regularly update the Voluntary Guidance to reflect lessons learned, new data, and stakeholder input as technology continues to be developed and refined.

For overall awareness and to ensure consistency in taxonomy usage, NHTSA adopted SAE International's Levels of Automation and other applicable terminology.[3]

NHTSA'S MISSION

Save lives, prevent injuries, and reduce economic costs due to road traffic crashes, through education, research, safety standards, and enforcement activity.

SECTION 1: VOLUNTARY GUIDANCE

SCOPE AND PURPOSE

Through this Voluntary Guidance, NHTSA is supporting entities that are designing ADSs for use on public roadways in the United States. This includes traditional vehicle manufacturers as well as other entities involved with manufacturing, designing, supplying, testing, selling, operating, or deploying ADSs, including equipment designers and suppliers; entities that outfit any vehicle with automated capabilities or equipment for testing, for commercial sale, and/or for use on public roadways; transit companies; automated fleet operators; "driverless" taxi companies; and any other individual or entity that offers services utilizing ADS technology (referred to collectively as "entities" or "industry").

This Voluntary Guidance applies to the design aspects of motor vehicles and motor vehicle equipment under NHTSA's jurisdiction, including low-speed vehicles, motorcycles, passenger vehicles, medium-duty vehicles, and heavy-duty CMVs such as large trucks and buses. These entities are subject to NHTSA's defect, recall, and enforcement authority.[4] For entities seeking to request regulatory action (e.g., petition for exemption or interpretation) from NHTSA, an informational resource is available on the Agency's website at www.nhtsa.gov/technology-innovation/automated-vehicles, along with other associated references and resources.

Interstate motor carrier operations and CMV drivers fall under the jurisdiction of FMCSA and are not within the scope of this Voluntary Guidance. Currently, per the Federal Motor Carrier Safety Regulations (FMCSRs), a trained commercial driver must be behind the wheel at all times, regardless of any automated driving technologies available on the CMV, unless a petition for a waiver or exemption has been granted. For more information regarding CMV operations and automated driving technologies, including guidance on FMCSA's petition process, see www.fmcsa.dot.gov.

This Voluntary Guidance focuses on vehicles that incorporate SAE Automation Levels 3 through 5 – Automated Driving Systems (ADSs). ADSs may include systems for which there is no human driver or for which the human driver can give control to the ADS and would not be expected to perform any driving-related tasks for a period of time.[5] It is an entity's responsibility to determine its system's automation level in conformity with SAE International's published definitions.

The purpose of this Voluntary Guidance is to help designers of ADSs analyze, identify, and resolve safety considerations prior to deployment using their own, industry, and other best practices. It outlines 12 safety elements, which the Agency believes represent the consensus across the industry, that are generally considered to be the most salient design aspects to consider and address when developing, testing, and deploying ADSs on public roadways. Within each safety design element, entities are encouraged to consider and document their use of industry standards, best practices, company policies, or other methods they have employed to provide for increased system safety in real-world conditions. The 12 safety design elements apply to both ADS original equipment and to replacement equipment or updates (including software updates/upgrades) to ADSs.

This Voluntary Guidance provides recommendations and suggestions for industry's consideration and discussion. This Guidance is entirely voluntary, with no compliance requirement or enforcement mechanism. The sole purpose of this Guidance is to support the industry as it develops best practices in the design, development, testing, and deployment of automated vehicle technologies.

NHTSA'S ENFORCEMENT AUTHORITY

Several States have sought clarification of NHTSA's enforcement authority with respect to ADSs. As DOT is asking States to maintain the delineation of Federal and State regulatory authority, NHTSA understands that States are looking for reassurance that the Federal Government has tools to keep their roadways safe.

NHTSA has broad enforcement authority to address existing and new automotive technologies and equipment. The Agency is commanded by Congress[6] to protect the safety of the driving public against unreasonable risks of harm that may arise because of the design, construction, or performance of a motor vehicle or motor vehicle equipment, and to mitigate risks of harm, including risks that may arise in connection with ADSs. Specifically, NHTSA's enforcement authority concerning safety-related defects in motor vehicles and motor vehicle equipment extends and applies equally to current and emerging ADSs. As NHTSA has always done, when evaluating new automotive technologies, it will be guided by its statutory mission, the laws it is obligated to enforce, and the benefits of the technology.

SECTION 1: VOLUNTARY GUIDANCE

SAE AUTOMATION LEVELS

Full Automation

0 — No Automation
Zero autonomy; the driver performs all driving tasks.

1 — Driver Assistance
Vehicle is controlled by the driver, but some driving assist features may be included in the vehicle design.

2 — Partial Automation
Vehicle has combined automated functions, like acceleration and steering, but the driver must remain engaged with the driving task and monitor the environment at all times.

3 — Conditional Automation
Driver is a necessity, but is not required to monitor the environment. The driver must be ready to take control of the vehicle at all times with notice.

4 — High Automation
The vehicle is capable of performing all driving functions under certain conditions. The driver may have the option to control the vehicle.

5 — Full Automation
The vehicle is capable of performing all driving functions under all conditions. The driver may have the option to control the vehicle.

ADS SAFETY ELEMENTS

1. System Safety

Entities are encouraged to follow a robust design and validation process based on a systems-engineering approach with the goal of designing ADSs free of unreasonable safety risks. The overall process should adopt and follow industry standards, such as the functional safety[7] process standard for road vehicles, and collectively cover the entire operational design domain (i.e., operating parameters and limitations) of the system. Entities are encouraged to adopt voluntary guidance, best practices, design principles, and standards developed by established and accredited standards-developing organizations (as applicable) such as the International Standards Organization (ISO) and SAE International, as well as standards and processes available from other industries such as aviation, space, and the military[8] and other applicable standards or internal company processes as they are relevant and applicable. See NHTSA's June 2016 report, *Assessment of Safety Standards for Automotive Electronic Control Systems*[9], which provides an evaluation of the strengths and limitations of such standards.

The design and validation process should also consider including a hazard analysis and safety risk assessment for ADSs, for the overall vehicle design into which it is being integrated, and when applicable, for the broader transportation ecosystem. Additionally, the process shall describe design redundancies and safety strategies for handling ADS malfunctions. Ideally, the process should place significant emphasis on software development, verification, and validation. The software development process is one that should be well-planned, well-controlled, and well-documented to detect and correct unexpected results from software updates. Thorough and measurable software testing should complement a structured and documented software development and change management process and should be part of each software version release. Industry is encouraged to monitor the evolution, implementation, and safety assessment of artificial intelligence and other relevant software technologies and algorithms to improve the effectiveness and safety of ADSs.

Design decisions should be linked to the assessed risks that could impact safety-critical system functionality. Design safety considerations should include design architecture, sensors, actuators, communication failure, potential software errors, reliability, potential inadequate control, undesirable control actions, potential collisions with environmental objects and other road users, potential collisions that could be caused by actions of an ADS, leaving the roadway, loss of traction or stability, and violation of traffic laws and deviations from normal (expected) driving practices.

All design decisions should be tested, validated, and verified as individual subsystems and as part of the entire vehicle architecture. Entities are encouraged to document the entire process; all actions, changes, design choices, analyses, associated testing, and data should be traceable and transparent.

SECTION 1: VOLUNTARY GUIDANCE

2. Operational Design Domain

Entities are encouraged to define and document the Operational Design Domain (ODD) for each ADS available on their vehicle(s) as tested or deployed for use on public roadways, as well as document the process and procedure for assessment, testing, and validation of ADS functionality with the prescribed ODD. The ODD should describe the specific conditions under which a given ADS or feature is intended to function. The ODD is the definition of where (such as what roadway types and speeds) and when (under what conditions, such as day/night, weather limits, etc.) an ADS is designed to operate.

The ODD would include the following information at a minimum to define each ADS's capability limits/boundaries:

- Roadway types (interstate, local, etc.) on which the ADS is intended to operate safely;
- Geographic area (city, mountain, desert, etc.);
- Speed range;
- Environmental conditions in which the ADS will operate (weather, daytime/nighttime, etc.); and
- Other domain constraints.

An ADS should be able to operate safely within the ODD for which it is designed. In situations where the ADS is outside of its defined ODD or in which conditions dynamically change to fall outside of the ADS's ODD, the vehicle should transition to a minimal risk condition.[10] For a Level 3 ADS, transitioning to a minimal risk condition could entail transitioning control to a receptive, fallback-ready user.[11] In cases the ADS does not have indications that the user is receptive and fallback-ready, the system should continue to mitigate manageable risks, which may include slowing the vehicle down or bringing the vehicle to a safe stop. To support the safe introduction of ADSs on public roadways and to speed deployment, the ODD concept provides the flexibility for entities to initially limit the complexity of broader driving challenges in a confined ODD.

3. Object and Event Detection and Response

Object and Event Detection and Response (OEDR)[12] refers to the detection by the driver or ADS of any circumstance that is relevant to the immediate driving task, as well as the implementation of the appropriate driver or system response to such circumstance. For the purposes of this Guidance, an ADS is responsible for performing OEDR while it is engaged and operating in its defined ODD.

Entities are encouraged to have a documented process for assessment, testing, and validation of their ADS's OEDR capabilities. When operating within its ODD, an ADS's OEDR functions are expected to be able to detect and respond to other vehicles (in and out of its travel path), pedestrians, bicyclists, animals, and objects that could affect safe operation of the vehicle.

An ADS's OEDR should also include the ability to address a wide variety of foreseeable encounters, including emergency vehicles, temporary work zones, and other unusual conditions (e.g., police manually directing traffic or other first responders or construction workers controlling traffic) that may impact the safe operation of an ADS.

Normal Driving

Entities are encouraged to have a documented process for the assessment, testing, and validation of a variety of behavioral competencies for their ADSs. Behavioral competency refers to the ability of an ADS to operate in the traffic conditions that it will regularly encounter, including keeping the vehicle in a lane, obeying traffic laws, following reasonable road etiquette, and responding to other vehicles or hazards.[13] While research conducted by California PATH[14] provided a set of minimum behavioral competencies for ADSs,[15] the full complement of behavioral competencies a particular ADS would be expected to demonstrate and routinely perform will depend upon the individual ADS, its ODD, and the designated fallback (minimal risk condition) method. Entities are encouraged to consider all known behavioral competencies in the design, test, and validation of their ADSs.

Crash Avoidance Capability – Hazards

Entities are encouraged to have a documented process for assessment, testing, and validation of their crash avoidance capabilities and design choices. Based on the ODD, an ADS should be able to address applicable pre-crash scenarios[16] that relate to control loss; crossing-path crashes; lane change/merge; head-on and opposite-direction travel; and rear-end, road departure, and low-speed situations such as backing and parking maneuvers.[17] Depending on the ODD, an ADS may be expected to handle many of the pre-crash scenarios that NHTSA has identified previously.[18]

The Federal Government wants to ensure it does not impede progress with unnecessary or unintended barriers to innovation. Safety remains the number one priority for U.S. DOT and is the specific focus of NHTSA.

SECTION 1: VOLUNTARY GUIDANCE

4. Fallback (Minimal Risk Condition)

Entities are encouraged to have a documented process for transitioning to a minimal risk condition when a problem is encountered or the ADS cannot operate safely. ADSs operating on the road should be capable of detecting that the ADS has malfunctioned, is operating in a degraded state, or is operating outside of the ODD. Furthermore, ADSs should be able to notify the human driver of such events in a way that enables the driver to regain proper control of the vehicle or allows the ADS to return to a minimal risk condition independently.

Fallback strategies should take into account that, despite laws and regulations to the contrary, human drivers may be inattentive, under the influence of alcohol or other substances, drowsy, or otherwise impaired.

Fallback actions are encouraged to be administered in a manner that will facilitate safe operation of the vehicle and minimize erratic driving behavior. Such fallback actions should also consider minimizing the effects of errors in human driver recognition and decision-making during and after transition to manual control.

In cases of higher automation in which a human driver may not be available, the ADS must be able to fallback into a minimal risk condition without the need for driver intervention.

A minimal risk condition will vary according to the type and extent of a given failure, but may include automatically bringing the vehicle to a safe stop, preferably outside of an active lane of traffic. Entities are encouraged to have a documented process for assessment, testing, and validation of their fallback approaches.

The purpose of this Voluntary Guidance is to help designers of ADSs analyze, identify, and resolve safety considerations prior to deployment using their own, industry, and other best practices. It outlines 12 safety elements, which the Agency believes represent the consensus across the industry, that are generally considered to be the most salient design aspects to consider and address when developing, testing, and deploying ADSs on public roadways.

5. Validation Methods

Given that the scope, technology, and capabilities vary widely for different automation functions, entities are encouraged to develop validation methods to appropriately mitigate the safety risks associated with their ADS approach. Tests should demonstrate the behavioral competencies an ADS would be expected to perform during normal operation, the ADS's performance during crash avoidance situations, and the performance of fallback strategies relevant to the ADS's ODD.

To demonstrate the expected performance of an ADS for deployment on public roads, test approaches may include a combination of simulation, test track, and on-road testing.

Prior to on-road testing, entities are encouraged to consider the extent to which simulation and track testing may be necessary. Testing may be performed by the entities themselves, but could also be performed by an independent third party.

Entities should continue working with NHTSA and industry standards organizations (SAE, International Organization for Standards [ISO], etc.) and others to develop and update tests that use innovative methods as well as to develop performance criteria for test facilities that intend to conduct validation tests.

SECTION 1: VOLUNTARY GUIDANCE

6. Human Machine Interface

Understanding the interaction between the vehicle and the driver, commonly referred to as "human machine interface" (HMI), has always played an important role in the automotive design process. New complexity is introduced to this interaction as ADSs take on driving functions, in part because in some cases the vehicle must be capable of accurately conveying information to the human driver regarding intentions and vehicle performance. This is particularly true for ADSs in which human drivers may be requested to perform any part of the driving task. For example, in a Level 3 vehicle, the driver always must be receptive to a request by the system to take back driving responsibilities. However, a driver's ability to do so is limited by their capacity to stay alert to the driving task and thus capable of quickly taking over control, while at the same time not performing the actual driving task until prompted by the vehicle. Entities are encouraged to consider whether it is reasonable and appropriate to incorporate driver engagement monitoring in cases where drivers could be involved in the driving task so as to assess driver awareness and readiness to perform the full driving task.

Entities are also encouraged to consider and document a process for the assessment, testing, and validation of the vehicle's HMI design. Considerations should be made for the human driver, operator, occupant(s), and external actors with whom the ADS may have interactions, including other vehicles (both traditional and those with ADSs), motorcyclists, bicyclists, and pedestrians. HMI design should also consider the need to communicate information regarding the ADS's state of operation relevant to the various interactions it may encounter and how this information should be communicated.

In vehicles that are anticipated not to have driver controls, entities are encouraged to design their HMI to accommodate people with disabilities (e.g., through visual, auditory, and haptic displays).[19]

In vehicles where an ADS may be intended to operate without a human driver or even any human occupant, the remote dispatcher or central control authority, if such an entity exists, should be able to know the status of the ADS at all times. Examples of these may include unoccupied SAE Automation Level 4 or 5 vehicles, automated delivery vehicles, last-mile special purpose ground drones, and automated maintenance vehicles.

Given the ongoing research and rapidly evolving nature of this field, entities are encouraged to consider and apply voluntary guidance, best practices, and design principles published by SAE International, ISO, NHTSA, the American National Standards Institute (ANSI), the International Commission on Illumination (CIE), and other relevant organizations, based upon the level of automation and expected level of driver engagement.

AT MINIMUM

An ADS should be capable of informing the human operator or occupant through various indicators that the ADS is:

- Functioning properly;
- Currently engaged in ADS mode;
- Currently "unavailable" for use;
- Experiencing a malfunction; and/or
- Requesting control transition from the ADS to the operator.

7. Vehicle Cybersecurity

Entities are encouraged to follow a robust product development process based on a systems engineering approach to minimize risks to safety, including those due to cybersecurity threats and vulnerabilities. This process should include a systematic and ongoing safety risk assessment for each ADS, the overall vehicle design into which it is being integrated, and when applicable, the broader transportation ecosystem.[20]

Entities are encouraged to design their ADSs following established best practices for cyber vehicle physical systems. Entities are encouraged to consider and incorporate voluntary guidance, best practices, and design principles published by National Institute of Standards and Technology (NIST[21]), NHTSA, SAE International, the Alliance of Automobile Manufacturers, the Association of Global Automakers, the Automotive Information Sharing and Analysis Center (Auto-ISAC),[22] and other relevant organizations, as appropriate.

NHTSA encourages entities to document how they incorporated vehicle cybersecurity considerations into ADSs, including all actions, changes, design choices, analyses, and associated testing, and ensure that data is traceable within a robust document version control environment.

Industry sharing of information on vehicle cybersecurity facilitates collaborative learning and helps prevent industry members from experiencing the same cyber vulnerabilities. Entities are encouraged to report to the Auto-ISAC all discovered incidents, exploits, threats and vulnerabilities from internal testing, consumer reporting, or external security research as soon as possible, regardless of membership. Entities are further encouraged to establish robust cyber incident response plans and employ a systems engineering approach that considers vehicle cybersecurity in the design process. Entities involved with ADSs should also consider adopting a coordinated vulnerability reporting/disclosure policy.

SECTION 1: VOLUNTARY GUIDANCE

8. Crashworthiness

Occupant Protection

Given that a mix of vehicles with ADSs and those without will be operating on public roadways for an extended period of time, entities still need to consider the possible scenario of another vehicle crashing into an ADS-equipped vehicle and how to best protect vehicle occupants in that situation. Regardless of whether the ADS is operating the vehicle or the vehicle is being driven by a human driver, the occupant protection system should maintain its intended performance level in the event of a crash.

Entities should consider incorporating information from the advanced sensing technologies needed for ADS operation into new occupant protection systems that provide enhanced protection to occupants of all ages and sizes. In addition to the seating configurations evaluated in current standards, entities are encouraged to evaluate and consider additional countermeasures that will protect all occupants in any alternative planned seating or interior configurations during use.[23]

Compatibility

Unoccupied vehicles equipped with ADSs should provide geometric and energy absorption crash compatibility with existing vehicles on the road.[24] ADSs intended for product or service delivery or other unoccupied use scenarios should consider appropriate vehicle crash compatibility given the potential for interactions with vulnerable road users and other vehicle types.

Entities are not required to submit a Voluntary Safety Self-Assessment, nor is there any mechanism to compel entities to do so. While these assessments are encouraged prior to testing and deployment, NHTSA does not require that entities provide disclosures nor are they required to delay testing or deployment. Assessments are not subject to Federal approval.

9. Post-Crash ADS Behavior

Entities engaging in testing or deployment should consider methods of returning ADSs to a safe state immediately after being involved in a crash. Depending upon the severity of the crash, actions such as shutting off the fuel pump, removing motive power, moving the vehicle to a safe position off the roadway (or safest place available), disengaging electrical power, and other actions that would assist the ADSs should be considered. If communications with an operations center, collision notification center, or vehicle communications technology exist, relevant data is encouraged to be communicated and shared to help reduce the harm resulting from the crash.

Additionally, entities are encouraged to have documentation available that facilitates the maintenance and repair of ADSs before they can be put back in service. Such documentation would likely identify the equipment and the processes necessary to ensure safe operation of the ADSs after repairs.

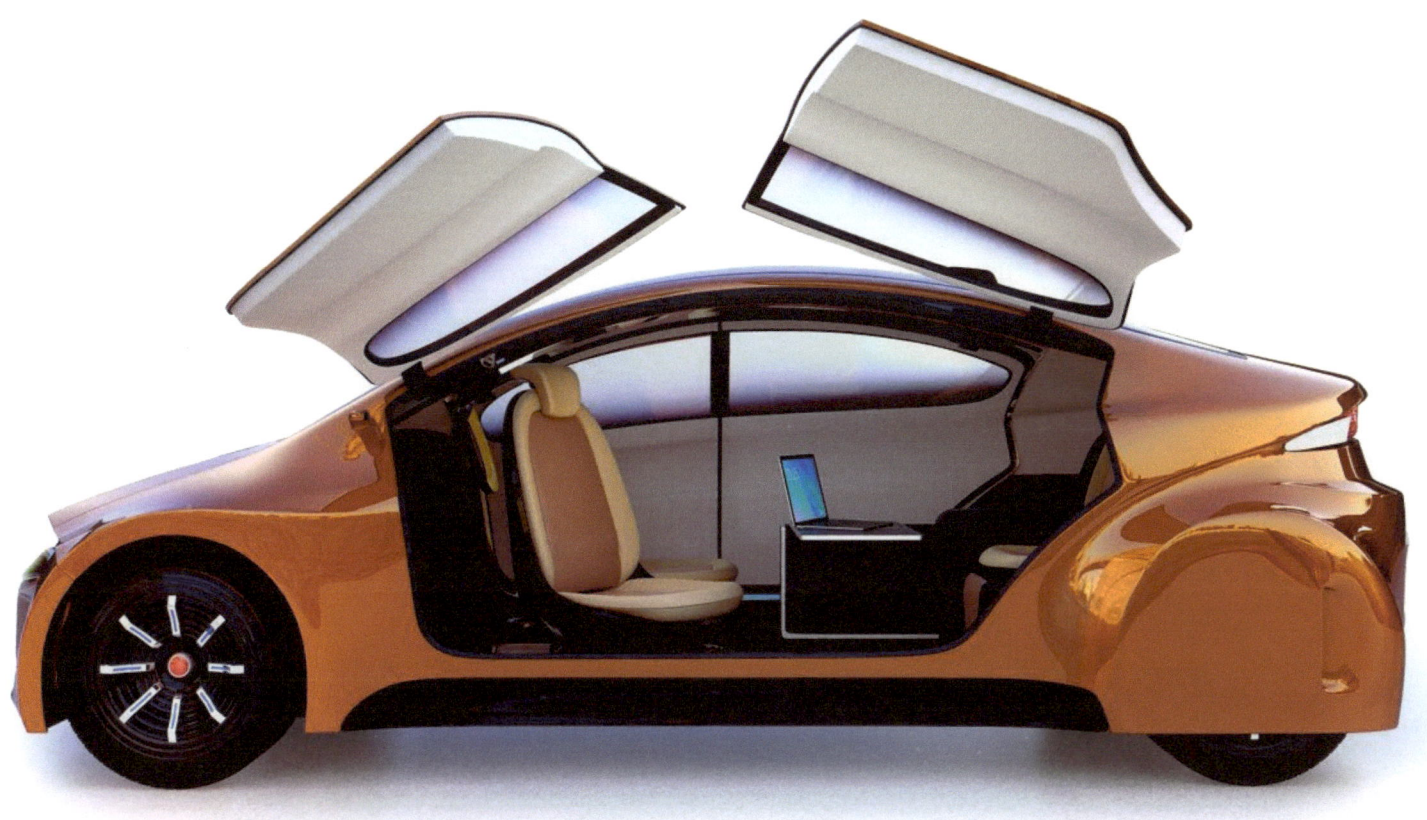

SECTION 1: VOLUNTARY GUIDANCE

10. Data Recording

Learning from crash data is a central component to the safety potential of ADSs. For example, the analysis of a crash involving a single ADS could lead to safety developments and subsequent prevention of that crash scenario in other ADSs. Paramount to this type of learning is proper crash reconstruction. Currently, no standard data elements exist for law enforcement, researchers, and others to use in determining why an ADS-enabled vehicle crashed. Therefore, entities engaging in testing or deployment are encouraged to establish a documented process for testing, validating, and collecting necessary data related to the occurrence of malfunctions, degradations, or failures in a way that can be used to establish the cause of any crash. Data should be collected for on-road testing and use, and entities are encouraged to adopt voluntary guidance, best practices, design principles, and standards issued by accredited standards developing organizations such as SAE International.[25] Likewise, these organizations are encouraged to be actively engaged in the discussion and regularly update standards as necessary and appropriate.

To promote a continual learning environment, entities engaging in testing or deployment should collect data associated with crashes involving: (1) fatal or nonfatal personal injury or (2) damage that requires towing, including damage that prevents a motor vehicle involved from being driven under its own power in its customary manner or damage that prevents a motor vehicle involved from being driven without resulting in further damage or causing a hazard to itself, other traffic elements, or the roadway.

For crash reconstruction purposes (including during testing), it is recommended that ADS data be stored, maintained, and readily available for retrieval as is current practice, including applicable privacy protections, for crash event data recorders.[26] Vehicles should record, at a minimum, all available information relevant to the crash, so that the circumstances of the crash can be reconstructed. These data should also contain the status of the ADS and whether the ADS or the human driver was in control of the vehicle leading up to, during, and immediately following a crash. Entities should have the technical and legal capability to share with government authorities the relevant recorded information as necessary for crash reconstruction purposes. Meanwhile, for consistency and to build public trust and acceptance, NHTSA will continue working with SAE International to begin the work necessary to establish uniform data elements for ADS crash reconstruction.

11. Consumer Education and Training

Education and training is imperative for increased safety during the deployment of ADSs.[27] Therefore, entities are encouraged to develop, document, and maintain employee, dealer, distributor, and consumer education and training programs to address the anticipated differences in the use and operation of ADSs from those of the conventional vehicles that the public owns and operates today.[28] Such programs should consider providing target users the necessary level of understanding to utilize these technologies properly, efficiently, and in the safest manner possible.

Entities, particularly those engaging in testing or deployment, should also ensure that their own staff, including their marketing and sales forces, understand the technology and can educate and train their dealers, distributors, and consumers.[29]

Consumer education programs are encouraged to cover topics such as ADSs' functional intent, operational parameters, system capabilities and limitations, engagement/disengagement methods, HMI, emergency fallback scenarios, operational design domain parameters (i.e., limitations), and mechanisms that could alter ADS behavior while in service. They should also include explicit information on what the ADS is capable and not capable of in an effort to minimize potential risks from user system abuse or misunderstanding.

As part of their education and training programs, ADS dealers and distributors should consider including an on-road or on-track experience demonstrating ADS operations and HMI functions prior to consumer release. Other innovative approaches (e.g., virtual reality or onboard vehicle systems) may also be considered, tested, and employed. These programs should be continually evaluated for their effectiveness and updated on a routine basis, incorporating feedback from dealers, customers, and other sources.

12. Federal, State, and Local Laws

Entities are also encouraged to document how they intend to account for all applicable Federal, State, and local laws in the design of their vehicles and ADSs. Based on the operational design domain(s), the development of ADSs should account for all governing traffic laws when operating in automated mode for the region of operation.[30] For testing purposes, an entity may rely on an ADS test driver or other mechanism to manage compliance with the applicable laws.

In certain safety-critical situations (such as having to cross double lines on the roadway to travel safely past a broken-down vehicle on the road) human drivers may temporarily violate certain State motor vehicle driving laws. It is expected that ADSs have the capability of handling such foreseeable events safely; entities are encouraged to have a documented process for independent assessment, testing, and validation of such plausible scenarios.

Given that laws and regulations will inevitably change over time, entities should consider developing processes to update and adapt ADSs to address new or revised legal requirements.

NHTSA encourages collaboration and communication between Federal, State, and local governments and the private sector as the technology evolves, and the Agency will continue to coordinate dialogue among all stakeholders. Collaboration is essential as our Nation embraces the many technological developments affecting our public roadways.

SECTION 1: VOLUNTARY GUIDANCE

VOLUNTARY SAFETY SELF-ASSESSMENT

Entities engaged in ADS testing and deployment may demonstrate how they address — via industry best practices, their own best practices, or other appropriate methods — the safety elements contained in the Voluntary Guidance by publishing a Voluntary Safety Self-Assessment. The Voluntary Safety Self-Assessment is intended to demonstrate to the public (particularly States and consumers) that entities are: (1) considering the safety aspects of ADSs; (2) communicating and collaborating with DOT; (3) encouraging the self-establishment of industry safety norms for ADSs; and (4) building public trust, acceptance, and confidence through transparent testing and deployment of ADSs. It also allows companies an opportunity to showcase their approach to safety, without needing to reveal proprietary intellectual property.

To facilitate this process and as an example of the type of information an entity might provide as part of its Voluntary Safety Self-Assessment, NHTSA has assembled an illustrative template for one of the safety elements within the Voluntary Guidance. This template is available on NHTSA's website. However, the information submitted could vary beyond the template when information is limited or unavailable (e.g., testing activities) or if the entity wishes to provide supplemental information.

Entities should ensure that Voluntary Safety Self-Assessments do not contain confidential business information (CBI), as it would be information available to the public. Entities will presumably wish to update these documents over time.

For each safety element laid out by the Voluntary Guidance, entities are encouraged to include an acknowledgment within the Voluntary Safety Self-Assessment that indicates one of the following:

- This safety element was considered during product development efforts for the subject feature; or
- This safety element is not applicable to the subject product development effort.

NHTSA envisions that the Voluntary Safety Self-Assessments would contain concise information on how entities are utilizing the Voluntary Guidance and/or their own processes to address applicable safety elements identified in the Voluntary Guidance. The Voluntary Safety Self-Assessment should not serve as an exhaustive recount of every action the entity took to address a particular safety element.

Entities are not required to submit a Voluntary Safety Self-Assessment, nor is there any mechanism to compel entities to do so. While these assessments are encouraged prior to testing and deployment, NHTSA does not require that entities provide submissions nor are they required to delay testing or deployment. Assessments are not subject to Federal approval.

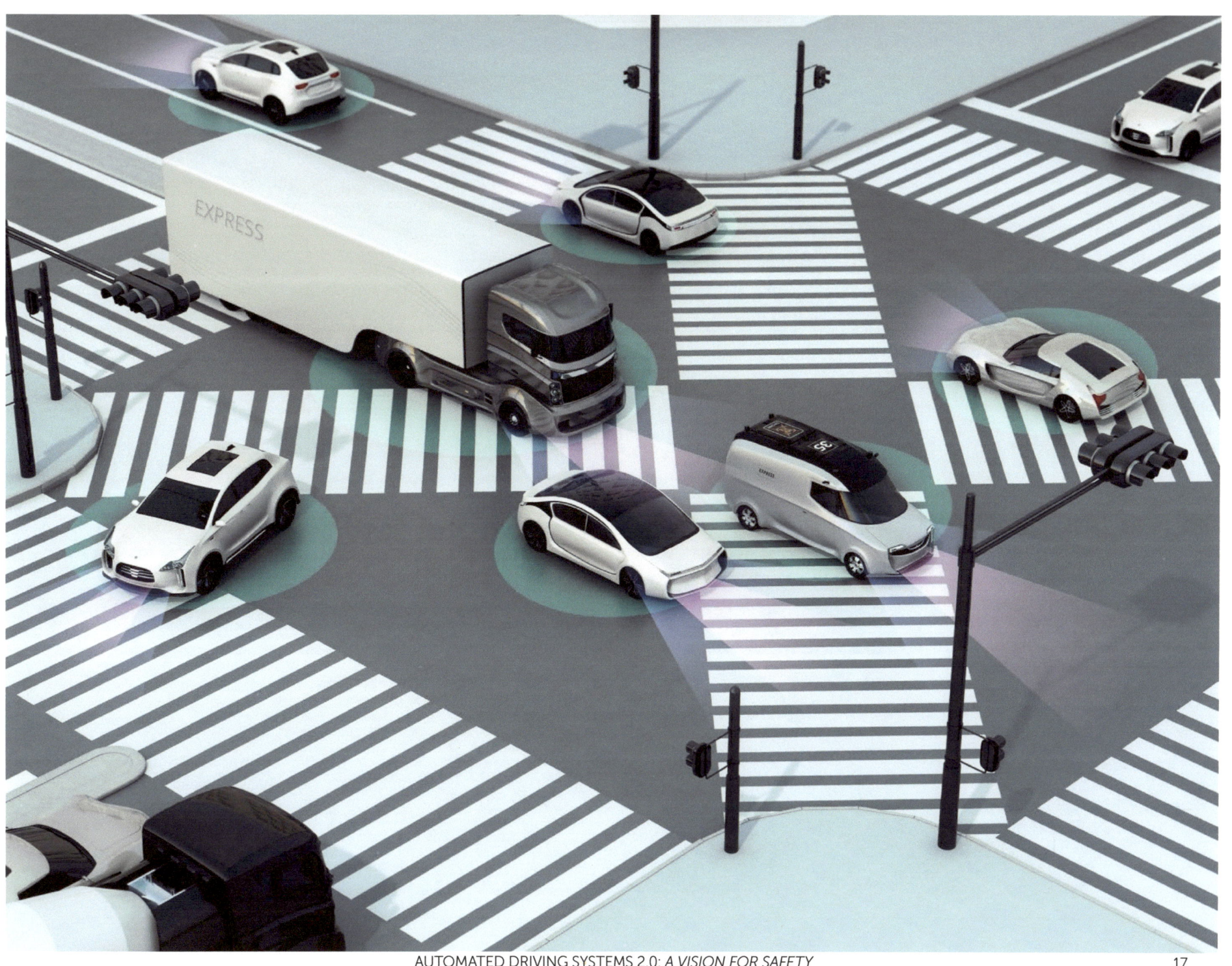

> **THE FEDERAL AND STATE ROLES**
>
> NHTSA strongly encourages States not to codify this Voluntary Guidance (that is, incorporate it into State statutes) as a legal requirement for any phases of development, testing, or deployment of ADSs. Allowing NHTSA alone to regulate the safety design and performance aspects of ADS technology will help avoid conflicting Federal and State laws and regulations that could impede deployment.

SECTION 2: TECHNICAL ASSISTANCE TO STATES

Best Practices for Legislatures Regarding Automated Driving Systems

OVERVIEW

The National Highway Traffic Safety Administration (NHTSA) of the U.S. Department of Transportation (DOT) is prepared to assist with challenges that States face regarding the safe integration of SAE Level 3 and above Automated Driving Systems (ADSs) on public roads. Given that vehicles operating on public roads are subject to both Federal and State jurisdictions and States are beginning to regulate ADSs, NHTSA has developed this section. It is designed to clarify and delineate the Federal and State roles in the regulation of ADSs and lay out a framework that the States can use as they write their laws and regulations surrounding ADSs to ensure a consistent, unified national framework.

NHTSA is working to bring ADSs safely onto the Nation's roadways in a way that encourages ADS entities (manufacturers, suppliers, transit operators, automated fleet operators, or any entity that offers services utilizing ADSs), consumer advocacy organizations, State legislatures, and other interested parties to work together in a shared environment. As the technology grows and the horizon of ADS changes rapidly, it is essential for each of these entities and interested parties to exercise due diligence in staying ahead of activity in a proactive—rather than reactive—manner.

States have begun to propose and pass legislation concerning ADSs. Public comments to NHTSA suggest that these proposals present several disparate approaches for adding and amending State authority over ADSs. Public comments and some State officials have asked NHTSA to provide guidance (and eventually regulations) that would support a more national approach to testing and deploying ADSs.

Further, in a prior collaborative effort between States and the Federal Government, NHTSA entered a 2-year cooperative agreement (beginning in September 2014) with the American Association of Motor Vehicle Administrators (AAMVA) under which the Autonomous Vehicle Best Practices Working Group was created. The working group was chartered to organize and share information related to the development, design, testing, use, and regulation of ADSs and other emerging vehicle technology. Based on the working group's research, a report is currently being developed to assist jurisdictions in enhancing their current ADS regulations or considering developing new legislation.[31] The goal of the report is to promote uniformity amongst jurisdictions and provide a baseline safety approach to possible challenges to the regulation of ADS sand testing the drivers who operate them.

Coinciding with the development of AAMVA's report, NHTSA has continued to work with State stakeholders including the National Conference of State Legislatures (NCSL) and the Governors Highway Safety Association (GHSA) to identify emerging challenges in the integration of ADSs and conventional motor vehicles.

Based on public input and the Agency's ongoing work with partners such as NCSL, GHSA, and AAMVA, NHTSA offers these Best Practices and specific legal components States should consider as we all work toward the shared goal of advancing safe ADS integration. The objective is to assist States in developing ADS laws, if desired, and creating consistency in ADS regulation across the country.

While technology is evolving and new State legislative language is still being drafted and reviewed, States can proactively evaluate current laws and regulations so as not to unintentionally create barriers to ADS operation, such as a requirement that a driver have at least one hand on the steering wheel at all times.

SECTION 2: TECHNICAL ASSISTANCE TO STATES

NHTSA encourages States to review others' draft ADS policies and legislation and work toward consistency. The goal of State policies in this realm need not be uniformity or identical laws and regulations across all States. Rather, the aim should be sufficient consistency of laws and policies to promote innovation and the swift, widespread, safe integration of ADSs.

States are encouraged to maintain a good state of infrastructure design, operation, and maintenance that supports ADS deployment and to adhere to the Manual on Uniform Traffic Control Devices (MUTCD), the existing national standard for traffic control devices as required by law. For example, items that may be considered a low priority now because of the presence of a human driver may be considered a higher priority as vehicle systems begin to rely more on machine vision and other techniques to detect where they are in a given lane. In addition, States are urged to continue to work with the Federal Highway Administration (FHWA) and the American Association of State Highway and Transportation Officials (AASHTO)[32] to support uniformity and consensus in infrastructure standards setting. This will support the safe operation of ADSs and ensure the safety of human drivers, who will continue to operate vehicles on the roads for years to come.

FEDERAL AND STATE REGULATORY ROLES

In consideration of State activity regarding ADSs, as well as NHTSA's activity at the Federal level, it is important to delineate Federal and State regulatory responsibility for motor vehicle operation.

These general areas of responsibility should remain largely unchanged for ADSs. NHTSA is responsible for regulating motor vehicles and motor vehicle equipment, and States are responsible for regulating the human driver and most other aspects of motor vehicle operation.

Further DOT involvement includes safety, evaluation, planning, and maintenance of the Nation's infrastructure through FHWA as well as regulation of the safe operation of interstate motor carriers and commercial vehicle drivers, along with registration and insurance requirements through the Federal Motor Carrier Safety Administration (FMCSA).

DOT strongly encourages States to allow DOT alone to regulate the safety design and performance aspects of ADS technology. If a State does pursue ADS performance-related regulations, that State should consult with NHTSA.

NHTSA'S RESPONSIBILITIES	STATES' RESPONSIBILITIES
• Setting Federal Motor Vehicle Safety Standards (FMVSSs) for new motor vehicles and motor vehicle equipment (with which manufacturers must certify compliance before they sell their vehicles)[33] • Enforcing compliance with FMVSSs • Investigating and managing the recall and remedy of noncompliances and safety-related motor vehicle defects nationwide • Communicating with and educating the public about motor vehicle safety issues	• Licensing human drivers and registering motor vehicles in their jurisdictions • Enacting and enforcing traffic laws and regulations • Conducting safety inspections, where States choose to do so • Regulating motor vehicle insurance and liability

BEST PRACTICES FOR LEGISLATURES

As States act to ensure the safety of road users in their jurisdictions, NHTSA continually monitors and reviews language to stay informed on State legislation. In reviewing draft State legislation, the Agency has identified common components and has highlighted significant elements regarding ADSs that States should consider including in legislation. As such, NHTSA recommends the following safety-related best practices when crafting legislation for ADSs:

- **Provide a "technology-neutral" environment.**

  States should not place unnecessary burdens on competition and innovation by limiting ADS testing or deployment to motor vehicle manufacturers only. For example, no data suggests that experience in vehicle manufacturing is an indicator of the ability to safely test or deploy vehicle technology. All entities that meet Federal and State law prerequisites for testing or deployment should have the ability to operate in the State.

- **Provide licensing and registration procedures.**

  States are responsible for driver licensing and vehicle registration procedures. To support these efforts, NHTSA recommends defining "motor vehicle" under ADS laws to include any vehicle operating on the roads and highways of the State; licensing ADS entities and test operators for ADSs; and registering all vehicles equipped with ADSs and establishing proof of financial responsibility requirements in the form of surety bonds or self-insurance. These efforts provide States with the same information as that collected for conventional motor vehicles and improve State recordkeeping for ADS operation.

- **Provide reporting and communications methods for Public Safety Officials.**

  States can take steps to monitor safe ADS operation through reporting and communications mechanisms so that entities can coordinate with public safety agencies. The safety of public safety officials, other road users, and ADS passengers will be improved with greater understanding of the technology, capabilities, and functioning environment. States should develop procedures for entities to report crashes and other roadway incidents involving ADSs to law enforcement and first responders.

- **Review traffic laws and regulations that may serve as barriers to operation of ADSs.**

  States should review their vehicle codes, applicable traffic laws, and similar items to determine if there are unnecessary regulatory barriers that would prevent the testing and deployment of ADSs on public roads. For example, some States require a human operator to have one hand on the steering wheel at all times — a law that would pose a barrier to Level 3 through Level 5 ADSs.

BEST PRACTICES FOR STATE HIGHWAY SAFETY OFFICIALS

States have a general responsibility to reduce traffic crashes and the resulting deaths, injuries, and property damage for all road users in their jurisdictions. States use this authority to establish and maintain highway safety programs addressing: driver education and testing; licensing; pedestrian safety; law enforcement; vehicle registration and inspection; traffic control; highway design and maintenance; crash prevention, investigation, and recordkeeping; and emergency services. This includes any legal components States may wish to consider upon drafting legislation on ADSs.

The following sections describe a framework for States looking for assistance in developing procedures and conditions for ADSs' introduction onto public roadways. NHTSA and AAMVA's collaborative partnership on a Model State Policy is the foundation of the following discussion; however, it has been upgraded to incorporate additional concerns of State stakeholders, the clarification of roles, and an emphasis on the States' consideration of the information—rather than a directive for action. NHTSA does not expect that States will necessarily need to create any new processes or requirements in order to support ADS activities. Instead, the references below are intended as guidance for those States that may be looking to incorporate ADSs into existing processes or requirements or States who are considering such processes or requirements.

1. **Administrative:** States may want to consider new oversight activities on an administrative level to support States' roles and activities as they relate to ADSs. NHTSA does not expect that States will need to create any particular new entity in order to support ADS activities, but States may decide to create some of these entities if the State determines that they will be useful. The references below are intended as examples of those that may be appropriate for participation.

   a. Consider identifying a lead agency responsible for deliberation of any ADS testing.

   b. Consider creating a jurisdictional ADS technology committee that is launched by the designated lead agency and includes representatives from the governor's office, the motor vehicle administration, the State department of transportation, the State law enforcement agency, the State Highway Safety Office, State office of information technology, State insurance regulator, the State office(s) representing the aging and disabled communities, toll authorities, trucking and bus authorities, and transit authorities.

   c. To encourage open communication, the designated lead agency may choose to inform the State automated safety technology committee of the requests from entities to test in their State and the status of the designated agency's response to companies.

   d. In an effort to implement a framework for policies and regulations, the designated lead agency could take steps to use or establish statutory authority. This preparation would involve examination of laws and regulations in order to address unnecessary barriers to ADS operation on public roadways.

   e. Consider developing an internal process to include an application for entities to test in their State.

   f. Consider establishing an internal process for issuing test ADS vehicle permits.

2. **Application for Entities to Test ADSs on Public Roadways:** For those States with an existing application process for test vehicles, the following are considerations for applications involving testing of an ADS on public roadways. It is recommended that the application for testing remain at the State level; however, if a State chooses to request applications at a local level, these considerations would carry to those jurisdictions.

a. States could request that an entity submit an application to the designated lead agency in each State in which it plans to test ADSs. A process should be considered for application submission in those situations in which multiple entities are involved in the testing of an ADS.

b. States could request the following information from entities to ensure accurate recordkeeping:

- Name, corporate physical and mailing addresses, in-State physical and mailing addresses (if applicable), and the program administrator/director's name and contact information;

- Identification of each ADS that will be used on public roadways by VIN, vehicle type, or other unique identifiers such as the year, make, and model; and

- Identification of each test operator, the operator's driver license number, and the State or country in which the operator is licensed.

c. Inclusion of the entity's safety and compliance plan for the ADS could provide increased safety assurance to the State.

d. Inclusion of evidence of the entity's ability to satisfy a judgment or judgments for damages for personal injury, death, or property damage caused by an ADS in the form of an instrument of insurance, a surety bond, or proof of self-insurance could provide increased safety assurance to the State.[34]

e. Inclusion of a summary of the training provided to the employees, contractors, or other users designated by the entity as test operators of the ADS could provide increased safety assurance to the State.

3. **Permission for Entities to Test ADSs on Public Roadways:**
For States that grant permission for testing of vehicles, the following are considerations for granting permission for ADS testing on public roadways. It is recommended that permission to test remain at the State level; however, State and local governments should coordinate. If a State chooses to request applications at a local level, these considerations would carry to those jurisdictions.

a. For greater public safety, it is recommended that a State's lead agency involve law enforcement agencies before responding to the application for testing from the entity.

b. It would be appropriate to suspend permission to test if the entity fails to comply with the State insurance or driver requirements.

c. It would be appropriate for the lead agency to request additional information or require an entity to modify its application before granting approval.

d. If a State requires an application, it should consider notification to the entity indicating permission to test that ADS in the State. A State may choose to request that entity's test vehicles carry a copy of proof of permission to test that ADS in those vehicles.

4. **Specific Considerations for ADS Test Drivers and Operations:** Considerations for States providing access for test-ADSs as they are operated under designated circumstances and with entity-based operators.

    a. If a State is concerned about the training of an ADS test driver, the State could request a summary of the training provided to the test driver.

    b. For test vehicles, the test driver should follow all traffic rules and report crashes as appropriate for the State.

    c. States regulate human drivers. Licensed drivers are necessary to perform the driving functions for motor vehicles equipped with automated safety technologies that are less than fully automated (SAE Levels 3 and lower). A licensed driver has responsibility to operate the vehicle, monitor the operation, or be immediately available to perform the driving task when requested or the lower level automated system disengages.

    d. Fully automated vehicles are driven entirely by the vehicle itself and require no licensed human driver (SAE levels 4 and 5), at least in certain environments or under certain conditions.[35] The entire driving operation (under specified conditions) is performed by a motor vehicle automated system from origin to destination.

5. **Considerations for Registration and Titling:** Specific considerations regarding identification and records for ADS deployed for consumer use and operation.

    a. Consider identification of an ADS on the title and registration. This could apply to all ADSs or only those capable of operating without a human driver.

    b. Consider requiring notification of ADS upgrades if the vehicle has been significantly upgraded post-sale. Applicable State forms could be adjusted to reflect the upgrade.

6. **Working With Public Safety Officials:** General considerations as public safety officials begin to understand vehicles and needs.

    a. States could consider training public safety officials in conjunction with ADS deployments in their jurisdictions to improve understanding of ADS operation and potential interactions.

    b. Coordination among States would be beneficial for developing policies on human operator behaviors, as to monitor behavior changes—if any—in the presence of ADSs when the vehicle is in control.

7. **Liability and Insurance:** Initial considerations for State relegation of liability during an incident and insurance of the driver, entity, and/or ADS. These considerations may take time and broad discussion of incident scenarios, understanding of technology, and knowledge of how the ADSs are being used (personal use, rental, ride share, corporate, etc.). Additionally, determination of the operator of an ADS, in a given circumstance, may not necessarily determine liability for crashes involving the ADS.

    a. Begin to consider how to allocate liability among ADS owners, operators, passengers, manufacturers, and other entities when a crash occurs.

    b. For insurance purposes, determine who (owner, operator, passenger, manufacturer, other entity, etc.) must carry motor vehicle insurance.

    c. States could begin to consider rules and laws allocating tort liability.

CONCLUSION

Public trust and confidence in the evolution of ADSs has the potential to advance or inhibit the testing and deployment of ADSs on public roadways. NHTSA is committed to supporting the safety of these emerging and evolutionary technological advancements, which have the potential to significantly improve roadway safety. The Voluntary Guidance, highlighting the 12 priority safety elements, and its associated Voluntary Safety Self-Assessment offer public reassurance that safety remains NHTSA's top priority. The States' Best Practices section reinforces NHTSA's willingness to assist States with the challenges they face regarding ADSs now and in the pivotal years ahead.

This document will be updated periodically to reflect advances in technology, increased presence of ADSs on public roadways, and any regulatory action or statutory changes that could occur at both the Federal and State levels. In the meantime, the information provided herein serves to aid industry as it moves forward with testing and deploying ADSs and States with drafting legislation and developing plans and policies regarding ADSs. NHTSA encourages collaboration and communication between Federal, State, and local governments and the private sector as the technology evolves, and the Agency will continue to coordinate dialogue among all stakeholders. Collaboration is essential as our Nation embraces the many technological developments affecting our public roadways. Together, we can use lessons learned to make any necessary course corrections, to prevent or mitigate unintended consequences or safety risks, and to positively transform American mobility safely and efficiently.

RESOURCES

A central repository of associated references to this and other NHTSA ADS resources will be maintained at
www.nhtsa.gov/technology-innovation/automated-vehicles.

This includes an informational resource to support manufacturers and other entities interested in requesting regulatory action from NHTSA.

ENDNOTES

1. NHTSA acknowledges that Privacy and Ethical Considerations are also important elements for entities to deliberate. See www.nhtsa.gov/AVforIndustry for NHTSA's approach on each.

2. NHTSA completed the Paperwork Reduction Act (PRA) process and received clearance from the Office of Management and Budget (OMB) on the Federal Automated Vehicles Policy Voluntary Guidance's information collection through August 31, 2018, 81 FR 65709. However, pursuant to PRA, NHTSA is again seeking public comment on an updated Information Collection Request (ICR) that covers the information included in Automated Driving Systems: A Vision for Safety. The ICR identified in this document will not be effective until the ICR process is completed.

3. SAE International J3016, International Taxonomy and Definitions for Terms Related to Driving Automation Systems for On-Road Motor Vehicles (J3016:Sept 2016).

4. See, e.g., 49 U.S.C. §§ 30102(a)(8), 30116, 30120.

5. Parts of this Voluntary Guidance could be applied to any form of ADS.

6. The National Traffic and Motor Vehicle Safety Act, as amended ("Safety Act"), 49 U.S.C. 30101 et seq., provides the basis and framework for NHTSA's enforcement authority over motor vehicle and motor vehicle equipment defects and non-compliances with Federal Motor Vehicle Safety Standards (FMVSS).

7. Under ISO 26262 (Road Vehicles: Functional Safety), functional safety refers to the absence of unreasonable safety risks in cases of electrical and electronic failures.

8. For example, the U.S. Department of Defense standard practice on system safety, MIL-STD-882E. 11 May 2012. Available at www.system-safety.org/Documents/MIL-STD-882E.pdf.

9. See Van Eikema Hommes, Q.D. (2016, June). *Assessment of Safety Standards for Automotive Electronic Control Systems*. (Report No. Dot HS 812 285). Washington, DC: National Highway Traffic Safety Administration. Available at ntl.bts.gov/lib/59000/59300/59359/812285_ElectronicsReliabilityReport.pdf.

10. "Minimal risk condition" means low-risk operating condition that an automated driving system automatically resorts to either when a system fails or when the human driver fails to respond appropriately to a request to take over the dynamic driving task. See SAE International J3016, International Taxonomy and Definitions for Terms Related to Driving Automation Systems for On-Road Motor Vehicles (J3016:Sept2016).

11. "Fallback ready user" means the user of a vehicle equipped with an engaged ADS feature who is able to operate the vehicle and is receptive to ADS-issued requests to intervene and to evident dynamic driving task (DDT) performance-relevant system failures in the vehicle compelling him or her to perform the DDT fallback. See SAE International J3016, International Taxonomy and Definitions for Terms Related to Driving Automation Systems for On-Road Motor Vehicles (J3016:Sept2016).

12. See Automated Vehicle Research for Enhanced Safety: Final Report. Collision Avoidance Metrics Partnership, Automated Vehicle Research Consortium. June 2016. DTNH22-050H-01277. The report includes detailed functional descriptions for on-road driving automation levels and identifies potential objective test methods that could be used as a framework for evaluating emerging and future driving automation features. Available at www.noticeandcomment.com/Automated-Vehicle-Research-for-Enhanced-Safety-Final-Report-fn-459371.aspx.

13. See Nowakowski, C., et al., *Development of California Regulations to Govern the Testing and Operation of Automated Driving Systems*, California PATH Program, University of California, Berkeley, Nov. 14, 2014, pg. 10. Available at http://docs.trb.org/prp/15-2269.pdf.

14. California Partners for Advanced Transit and Highways (PATH) is a multidisciplinary research and development program of the University of California, Berkeley, with staff, faculty, and students from universities worldwide and cooperative projects with private industry, State and local agencies, and nonprofit institutions. See www.path.berkeley.edu.

15. Id., pgs. 10-11. California PATH's work described minimum behavioral competencies for automated vehicles as "necessary, but by no means sufficient, capabilities for public operation." Id. The document's full peer review is available at www.nspe.org/sites/default/files/resources/pdfs/Peer-Review-Report-IntgratedV2.pdf.

16. See Rau, P., Yanagisawa, M., and Najm, W. G., *Target Crash Population of Automated Vehicles*, available at www-esv.nhtsa.dot.gov/Proceedings/24/files/Session 21 Written.pdf.

17 See Najm, W. G., Smith, J. D., and Yanagisawa, M., "Pre-Crash Scenario Typology for Crash Avoidance Research," DOT HS 810 767, April 2007. Available at www.nhtsa.gov/gy-Final_PDF_Version_5-2-07.pdf.

18 Available at http://ntl.bts.gov/lib/55000/55400/55443/AVBenefitFrameworkFinalReport082615_Cover1.pdf.

19 Entities are encouraged to seek technical and engineering advice from members of the disabled community and otherwise engage with that community to develop designs informed by its needs and experiences.

20 Entities should insist that their suppliers build into their equipment robust cybersecurity features. Entities should also address cybersecurity, but they should not wait to receive equipment from a supplier before doing so.

21 www.nist.gov/cyberframework.

22 An Information Sharing and Analysis Center (ISAC) is a trusted, sector specific entity that can provide a 24-hour-per-day 7-day-per-week secure operating capability that establishes the coordination, information sharing, and intelligence requirements for dealing with cybersecurity incidents, threats, and vulnerabilities. See McCarthy, C., Harnett, K., Carter, A., and Hatipoglu, C. (2014, October). *Assessment of the information sharing and analysis center model* (Report No. DOT HS 812 076). Washington, DC: National Highway Traffic Safety Administration.

23 The tools to demonstrate such due care need not be limited to physical testing but also could include virtual tests with vehicle and human body models.

24 In 2003, as part of a voluntary agreement on crash compatibility, the Alliance of Automobile Manufacturers agreed to a geometric compatibility commitment which would provide for alignment of primary energy absorbing structures among vehicles. The European Union recently introduced a new frontal crash test that also requires geometric load distribution similar to the Alliance voluntary agreement.

25 The collection, recording, storage, auditing, and deconstruction of data recorded by an entity must be in strict accordance with the entity's consumer privacy and security agreements and notices, as well as any applicable legal requirements.

26 See 49 CFR Part 563, Event Data Recorders. Available at www.gpo.gov/fdsys/pkg/CFR-2016-title49-vol6/xml/CFR-2016-title49-vol6-part563.xml.

27 Not applicable to ADS testing.

28 The training and education programs recommended here are intended to complement and augment driver training and education programs run by States that retain the primary responsibility for training, testing, and licensing human drivers.

29 Such training and education programs for employees, dealers, distributors, and consumers may be administered by an entity other than the direct employer, manufacturer, or other applicable entity.

30 Traffic laws vary from State to State (and even city to city); ADSs should be able to follow all laws that apply to the applicable operational design domain. This includes speed limits, traffic control devices, one-way streets, access restrictions (crosswalks, bike lanes), U-turns, right-on-red situations, metering ramps, and other traffic circumstances and situations.

31 Future updates to AAMVA's guide may integrate commercial vehicle ADS operational aspects brought forth by the Commercial Vehicle Safety Alliance (CVSA).

32 AASHTO is an international leader in setting technical standards for all phases of highway system development. Standards are issued for design, construction of highways and bridges, materials, and many other technical areas. See www.transportation.org/home/organization/.

33 NHTSA does not expressly regulate motor vehicle (or motor vehicle equipment) in-use performance after first sale. However, because the FMVSSs apply to the vehicle or equipment when first manufactured and because taking a vehicle or piece of equipment out of compliance with an applicable standard can be a violation of the Safety Act, the influence of the FMVSSs extends throughout the life of the vehicle even if NHTSA is not directly regulating it. At the same time, States have the authority to regulate a vehicle's in-use performance (through safety inspection laws), but as the text here states, State regulations cannot conflict with applicable FMVSSs. Additionally, NHTSA continues to have broad enforcement authority to evaluate and address safety risks as they arise.

34 AAMVA experts recommended a minimum insurance requirement of $5 million; however, that is subject to State considerations.

35 Some vehicles may be capable of being entirely "driven" either by the vehicle itself or by a human driver. For such dual-capable vehicles, the States would have jurisdiction to regulate (license, etc.) the human driver.

DOT HS 812 442
September 2017

www.ingramcontent.com/pod-product-compliance
Lightning Source LLC
Chambersburg PA
CBHW040452220526
45473CB00004B/1608